What D.E.W. You Think Happened In Hawaii?

Copyright © 2023 Wayne McRoy

All Rights Reserved

ISBN: 9798867397524

You're listening to The Alchemical Tech Revolution, and I am your host, Wayne McRoy.

Good evening, everyone.

Tonight, what do you think happened in Hawaii?

It's a question I think that needs to be asked, and I think answers need to be forthcoming.

1:21

Of course, they will always, always strongly deny

That this could have been done on purpose, these fires set on purpose.

We've seen this before in California.

We've seen this back in the year 2020 in Australia.

1:42

These places have been burned to the ground and of course it's all about a rebuilding effort after the fact here.

So there's been a lot of talk

About the possible use of directed energy weapons, now directed energy weapons, do they really exist?

2:01

Well, tonight we're going to answer that question and I don't think too many of you will be shocked at what the answer is, honestly here, but could this be what happened in Hawaii?

2:18

I think it seems feasible.

Of course, it's hard to tell much of anything these days.

In the early hours of an event like this, it's usually better to wait and see how it shakes out.

But it seems to me there's an awful lot of evidence to suggest that this has been planned in some way, shape or form.

2:36

It seems to me that even though we are surrounded by ineffective, doddling idiots

within government, even though that may be the case.

2:51

This is egregious.

What happened there?

The mismanagement is off the chain.

This can only be achieved on purpose,

this type of mismanagement, a total lack of common sense.

You want to talk about idiocracy, folks?

3:07

Well this is what happened in response to this.

If you want to be, or play, the devil's advocate,

And say that this was just government mismanagement and miscommunication and that kind of thing.

That kind of reasoning only goes so far.

If you're the guy that can control access to the water to put out these fires and you withhold access of that to these communities and just let them burn,

3:34

That, my friends, goes beyond incompetence.

That's goes beyond gross negligence.

That is criminal.

And that is one of the aspects of this that has happened, goes way beyond just simply mismanaging the response to this.

3:51

This is criminal, criminal.

Another aspect of this that really bothers me that I had seen is that the schools closed down and sent the children home with no adults present in their homes.

4:14

And many of these people allegedly died within the fires.

I have here an article I'll look at.

This article is from the Daily Mail and the headline is Maui Boy 7 is found burned to death in car as local lawmaker says she fears hundreds of children may be dead after Power Cut kept them home from school on the day of the inferno.

4:41

So the power was turned off on that day, so the schools were closed.

So they sent the kids home and the kids had nowhere to go.

No adult supervision.

So it says here, fears are growing that a large number of the Maui wildfire victims are children.

4:59

Many children were at home on the day of the fires after power cuts on the island closed schools, prompting fears some may not have escaped the flames.

And when this article was written, the death toll was currently 111.

But as of 47 minutes ago, according to what I've seen here, the death toll now stands at 114.

5:20

So it says here the death toll is currently 111.

But a lawmaker said she expects the final figure will be in the hundreds.

So we see here that it would appear that there were children burned alive in this fire due to

5:40

The mismanagement of the response to these events, that's if you believe the official story and narrative.

I don't know if we're being handed a bill of goods, folks can't say that for sure.

5:58

But what I do know is this has all of the hallmarks of an occult ritual.

An occult ritual.

You see the people who perform these occult rituals or plan them in this way, they call themselves the Philosophers of Fire, and they believe that through fire, nature is perfectly renewed.

6:25

This is why they burn their enemies.

There's also an occult connotation to that that.

If they properly burn their enemies in effigy, then the souls of their enemy will not come back to haunt them.

6:44

This is what they teach in some of the esoteric occult fraternities about these things.

That is why they burn their enemies.

That's why this is a favored tactic of these elitist scumbags, these dark occultists who run things.

7:00

They like to burn people.

That's what happened in Waco, TX.

That's what happened.

Well, you know, all the rest, all the places where fires were part of the demise of various people.

7:18

You know you've seen it before.

Not only that, you throw into the mix, the lifetime actor who managed the whole Las Vegas event where Jason Aldean performed,

was involved directly in this Hawaiian response here as well, part of the response team, So that should tell you something about that as well.

7:43

And the fact that Jason Aldean was just recently in news leading up to this event should be a tell to everybody there as well.

You see, these things just aren't random happenstances.

The people in charge here, these dark occultists who run things, would like you to believe, like

8:00

the average person would like to believe, all of this stuff just occurs,

It's coincidence.

There's none of this that goes on.

There's no preplanning.

There's no occultic side to it.

Well, most certainly there is.

There is an occult side to all of this.

8:17

There always is.

As much as I would like that to not be the case, it's always there folks.

It's always lurking in the background.

And all you have to do is look back and read the synchromystic metadata out there floating around in the ether to see what has been leading up to this.

8:37

And oftentimes, the nature of this metadata is such that you don't see it until after the event has occurred, and then you can go back and look and see it.

This is called revelation of the method.

That's what this is, so you can know.

8:52

That these dark occultists at the top of the power structure, they have their fingerprints all over this.

But the problem with these fingerprints is they're plausibly deniable, because most people can't accept the fact that there's an occult side to all of this and that the preplanning has been out there floating around for a long time leading up to this.

9:17

All you got to do is look at the Mountain Dew can.

A flavor they came out with in 2020.

Another banner year for the social programming of the masses, Social engineering on a grand scale.

9:37

That flavor is called Maui Burst.

Isn't that intriguing that they had a flavor of Mountain Dew called Maui Burst?

The symbology's been there, folks.

It's been out there.

You've probably seen a lot of it and not realized it, and you wouldn't realize it until it's pointed out to you.

9:57

Another one.

Go back and watch the Disney film Moana, centered around the Hawaiian Islands and the Hawaiian culture and their mythology.

Always the mythological aspect tied to this stuff as well, the archetype.

If you watch that film, you'll notice that at the end what one of the main premises of the film is.

10:21

The lava God, I don't remember

this lava monster's name particularly, comes alive and burns the island, and the only way for the island to be saved is for Moana to return this artifact to the spirit of the island, and then it can be reborn and renewed.

10:41

And one of the main antagonists/protagonists in that film is a demigod named Maui.

And Maui meets his destruction and demise at the hands,

well, nearly meets his destruction and demise at the hands of this fire monster at the end.

11:00

But of course he is saved and renewed in the rebirth process.

The regreening of the island.

The island being reborn and reimbued with life because of the returning of this artifact.

Which I think was called The Eye of Tafiti or Fahiti or some such thing, Tafiti in the film.

11:23

I haven't watched it in a while, I don't quite remember, but I do have a lot of children.

So that gives me an excuse to watch some of these films and I sit there and I pick out the symbology in them.

But of course you wouldn't recognize that at the time of its release because this wasn't a thing yet.

11:41

But here we are.

So now you have the esoteric

symbol of the All-seeing Eye present there in the rebirth of this now burned, destroyed land.

And of course this harkens back to the mythological aspect of the Phoenix. Through fire,

12:00

All nature is perfectly renewed.

All things are perfectly renewed.

That is the motto of the philosophers of fire.

And of course, there's a rebuilding campaign that's been suggested here

for this whole thing.

So the evidence there stands out.

12:20

As far as the occult side of this, it's definitely been circulating for a while.

The build up for this, the synchromystic metadata as I like to call it, was certainly out there pointing the way towards this.

And now, with hindsight being 2020, not coincidentally enough, we can see

12:43

these synchromystic connections and of course it's all about pattern recognition as well.

So this is a precursor event for something bigger to come.

In my estimation, sometime this fall, probably somewhere around say September 29ish.

Now that's a total guess, total guess based upon some things that I've seen.

13:04

I'm not making a prediction, but I do know

the way they like to leverage certain energetic principles in ways and September 29th is the next big date in the way that they've outlined this year of 2023 from the get go.

13:24

And how have they done that?

Well, you got to go back and look at the Chinese New Year celebration and all the symbology associated with that.

The Year of the Rabbit, Chinese New Year.

13:40

Their next big celebration will be September 29th.

So I think there's going to be some type of an event that occurs around that time where they're going to try to leverage off of that energetic principle in some way.

That's a guess.

Like I said, it's conjecture on my part at this point, but we'll see what happens.

14:01

Based upon how I know that these people operate,

I could certainly see them, perhaps trying something like that.

And maybe even the mere mention of that here is enough for them to not do something if they had something planned.

14:18

That's my hope in putting that out there.

So if something doesn't come to fruition, I'll be most happy about that.

Let's put it that way.

And I don't know whether to expect something or not.

Like I said, total conjecture on my part.

14:34

Total guesswork here, but I suspect,

That probably around that time we'll see them try the next phase of their operation in this year of the rabbit, the empowerment of the rabbit as the trickster in this way.

14:51

And of course they're ramping up all the same tropes that they were back in 2020.

I see that as being a banner year here for them.

It's pattern recognition.

Go back and look at the patterns in 2020.

The year started with Australia on fire.

15:10

Then we began to see some other things around that same time.

In late 2019, there was the UFO revelations.

And now this year we have UFO revelations just prior to this fire.

Then we had the big fire in Australia in 2020.

15:29

Then all of a sudden these monoliths began to appear around the world.

You remember all of that?

The harkening back to those monoliths, well, something similar began to happen where a totem pole began to just mysteriously over appeared overnight Someplace in the UK, I had seen the article and they were drawing attention to that.

15:55

The mainstream news for some reason.

And it's all pattern recognition, folks.

So it's the same kind of things that were going on.

So shortly after that happened back in 2020, we had on 3/11/2020, we experienced the lockdowns and all the COVID measures kicked into play and the world changed in a heartbeat that day.

16:21

The whole image of man and his place in this world changed overnight with that event.

And people capitulated to the nonsense.

And they have learned to regret that.

Now, three years later, over three years, it's been hell.

16:43

And now I've been hearing that in Canada they began to pass some very draconian laws.

They plan on keeping COVID measures in place for churches, specifically for churches in Canada.

17:01

That's right, you can go check.

I did go back and I checked this.

It's legit.

Their Supreme Court in Canada upheld COVID measures being essential for church services in Canada even today, three years after the fact.

17:25

So these are concerning things.

They're gearing up for something else, and this is one of their calling cards for that.

The massive fire.

We also had those massive fires in California, where in California? PARADISE.

California and now Hawaii, a paradise again.

17:45

So we've got Fire in paradise.

Fire in paradise.

That's a harbinger, folks.

For these folks, that's how they see it.

This harkens

back to the idea of burning it all down to build it anew.

18:05

Paradise representing the garden.

The garden of Eden.

Nature.

The natural order.

Their plan?

To completely replace nature with something totally artificial.

By burning it down, destroying it and rebuilding it, it'll be the rebirth.

18:21

The Phoenix.

You see, man thinks he can do it better than the creator.

That's always been the case that harkens back to the old tale in the Garden of Eden once again, and it's the hubris involved with this, absolute hubris.

18:43

So that being the case,

We're talking about the esoteric side of this, the occult side.

Is there any physical reality to directed energy weapons?

18:59

And this has been the speculation by many people, and it would seem that there might be some reason for people to make that conjecture looking at some of the evidence that we have seen.

Cars generally don't melt.

19:15

Metal generally doesn't melt in a forest fire.

A forest fire that generally doesn't touch the trees,

Isn't that strange?

Again, we have that same thing going on.

I guess you know, that's because trees have water in them,

I guess that's the official explanation, but that doesn't make sense because people have water in them and people certainly burn, don't they?

19:35

Houses.

Buildings burned.

Infrastructure burned.

But not some of the trees.

Seems to me that this was a perfect storm of events that had to coalesce together perfectly in order for this fire to spread how it did.

19:58

So what's the chances that this was mismanaged so badly in this way that this allowed this to happen?

That some, oh I don't know, accidental event, like maybe a downed power line,

Sparked the fire that started and then the gusting winds and the storms were enough to make that fire spread.

20:20

So then they turned off the power because of the wind and the storms, and this allowed people to be trapped, and they blocked off the roadways, keeping people in from getting to the coast or getting out of town to avoid panic, blocked off the roads, turned off the power, and then they cut off the water supply.

20:58

Do those sound like accidental occurrences?

Does that just sound like, oh, miscommunication, mismanagement?

And not only that, they didn't sound the alarm.

They have a storm, a tsunami alarm that in the case of an emergency, they are supposed to sound.

21:18

Now the idiot who was supposed to be in charge of that claims that the reason they didn't do that is because they didn't want people to run back inland into the fires.

If you're idiotic enough to run towards a fire, if you hear an alarm go off and you go outside and you see massive fire towards the center of the island, are you running there or are you running to the beach?

21:44

That is a poor excuse.

But these are the kinds of excuses they give.

So is this really incompetence on that level, or is this mismanagement on that level?

Or were these things planned to go in this way?

22:04

Or I'll give you a third option,

Is it all a bill of goods we're being sold and none of this stuff occurred as reported like this?

But this is what we're handed to create outrage from people, to keep us focused on this event when something bigger is on the horizon.

22:24

Is this really how it went down?

Did people really die?

And I won't go ahead and claim nobody died there.

It's certainly feasible that perhaps they did, and that it's even a greater tragedy if people really did die there.

But the response is part of the reason for the death toll.

22:44

If that's the case, and you have to wonder, was there some planning involved in this?

Is this typically how people respond?

With no common sense to an emergency situation, would you really?

23:00

If you're in charge of the water supply, there's massive fires all over.

"I better turn off the water".

Really.

Are you an idiot?

Let's block the roads and keep people from leaving town while these massive fires spread and they have no way to fight the fire because they don't have any fricking water.

23:29

They don't have any help.

They are unable to leave.

This is either gross incompetence on a massive scale, and that says something about our government, if that's the case, or, it was planned to go this way or the third option, none of it really occurred this way that's being reported, but we're handed this just to generate outrage,

24:01

Keep us focused on this event.

Focus on the details.

What's going on?

What started the fire could have been military weapons, directed energy weapons, could have been maybe a spark from, I don't know, downed power lines, could have been an idiot left a campfire burning or something.

24:27

We don't really know what started this or how it escalated out of control.

But in this day and age when they can clearly, clearly control the weather if they want to, they allowed this to happen.

24:43

They allowed these conditions to persist.

They allowed this fire to burn.

They clearly had access to water that they could have utilized and maybe diminished the damage a little bit.

They could have helped people to escape.

24:59

They could have given people better notice or warning.

They could have sounded their sirens that they have for just such emergencies.

Instead, they did nothing except block the roads and send children home to die.

25:16

If you believe this narrative, I don't know what to make of it.

Only time will tell here what comes of all of it.

But the big question on everybody's minds, was it directed energy weapons?

Well, tonight we're going to look, we're going to look at the state of directed energy weapons.

25:40

What's real, what's not?

What do they have?

What's available and where are we going to go for this?

I'm going to go to the Rand Corporation, The Rand Corporation's Project Air Force to be precise.

25:58

A little document here in my vast archive that's called "Space Weapons, Earth Wars" that talks about just this thing.

This was written in 2002, published in 2003 with some addendums in 2003 prepared for the United States Air Force, Approved for Public Release Distribution Unlimited from Project Air Force Rand Corporation, available in the Library of Congress Publication Data Record.

26:28

And it says the resource reported here was sponsored by the United States Air Force under contract F4-9642-01-C-0003.

Further information may be obtained from the Strategic Planning Division, Directorate of Plans Headquarters, United States Air Force.

26:48

Rand is a nonprofit institution that helps improve policy and decision making through research and analysis.

Rand is a registered trademark.

Rand's publications do not necessarily reflect the opinions or policies of its research

27:05

sponsors. Copyright 2002 Rand Corporation All rights reserved.

Published by Rand Santa Monica, CA.

Arlington, VA, Pittsburgh, PA.

So let's read the summary of this paper.

27:26

Space weapons for terrestrial conflict have been the subject of intense debate twice in the modern history of space.

The first time at the beginning of the Cold War was over the possibility of bombardment satellites carrying nuclear weapons.

27:42

The second time, at the end of the Cold War was over the possibility of space based defenses against nuclear missiles.

Now well past the Cold War, the topic of space weapons seems headed again for public debate, this time based on ballistic missile defense.

28:00

National policy documents tactically include the development of advanced technology to improve ballistic missile defense options.

The latest space policy document from the Department of Defense written by a gentleman named Cohen from 1999 supports quote, Ballistic missile defense and force projection.

28:20

End Quote.

To this end, the United States is developing a space based laser technology which is approaching the demonstration phase.

Kind of pause for a moment here, folks.

Remember this was 2002, 22 years ago.

28:41

For these reasons, as well as the threat that space weapons could pose if developed by an adversary, it is time for public discussion of the subject.

This report does not present an argument either for or against space weapons.

28:56

What instead describes their attributes and sets out a common vocabulary for future discussions.

The report classifieds and compares these weapons and explains how they might be used.

It also explores ways in which the United States and other countries might decide to acquire them and the potential reaction of other countries if the United States or some other nation fielded such weapons.

29:21

The report dispels some of the myths regarding space weapons to help ensure that debates and discussions are more fact based.

Going to pause for a moment here folks.

So we're talking about laser weapons.

That's how they framed it up here.

29:39

But actually it goes way beyond just a laser based weapons.

The things they have now go way beyond what they're acknowledging here.

And in fact, the technologies they're acknowledging here publicly in 2002 were likely developed in the 1970s or 80s.

29:59

So that's what was being disclosed at that time.

Now think about that.

Keep that in mind when it comes to exotic weapons like these within the military industrial complex.

Remember, they keep most of their more advanced developments under the appropriation of the special access programs within the black budget community of the military industrial complex.

30:28

A lot of these locked away in corporate vaults, somewhere away from the prying eyes of Congress and the public, not subject to FOIA request, Freedom of Information Act request because it could be regarded as, quote-unquote proprietary information, so they don't have to disclose what they're working on.

30:54

And of course you have to have a special compartmentalized top secret clearance to work in one of these special access programs.

So some of these technologies don't largely get acknowledged publicly until well after their testing phase, usually 20 to 30 years.

31:14

So the thing we see disclosed as the most advanced in the public sector is at least 20 to 30 years old and has been developed with all its weaponized potentials within the confines of the military industrial complex.

31:33

Otherwise a report like this would still be top secret or very much sensitive, be classified to some degree.

But this is for public review, public release in 2002.

Let's go ahead and read on the next section here talks about space weapons compared.

31:52

It is important to understand that space based weapons generally include several distinct classes of weapons.

Going to pause for a moment here folks, before we go through these.

If you have a problem with the notion of space or the existence of space, keep in mind what they're referring to as space is what I would call low Earth orbit, which essentially is very high up in the sky, very high up in the sky.

32:24

So is it possible they have platforms up there?

I think it's demonstrable that they do.

How they work, I don't know.

There's some people that claim that all satellites are just balloons.

I don't know if that's the case.

But what I do know is we've been lied about as to what exactly is up there and how it's kept up there, how it moves around up there.

32:50

But make no mistake about it, there are things up there in the sky, that's demonstrable.

And could it be that they have classified weapon systems up there?

Oh, you bet your bottom dollar they do.

No doubt in my mind they're up there.

And also keep in mind that out there in the UFO research community for many years people have been talking about weaponized satellites.

33:15

In fact, one Mr. John Lear back in the 1990s on the regular would talk about how the military has 24 weaponized satellites up there that can destroy things from space with a laser like technology.

33:33

The very thing we're describing here, a DEW, directed energy weapon.

This was the claim made by John Lear, who worked for the CIA, by the way.

So I don't know, you could take that with a grain of salt.

Could be disinformation or maybe he knew something.

Maybe there were technologies up there, high up in the sky that can perform these feats.

33:54

Certainly I think it's feasible and I think it's demonstrable.

Now, maybe it's not, quote-unquote, satellites, Maybe it's just advanced aircraft or drones that are flying around up there that they can do this with.

In fact, we've seen these weapon systems on airplanes, jets, drones.

34:16

What weapons systems are we talking about?

Well, here the Rand Corporation breaks down these different weapons, the space based weapons.

The first one is directed energy weapons.

The second one is kinetic energy weapons against missile targets.

34:32

The third one is kinetic energy weapons against surface targets and the 4th one is space based conventional weapons against surface targets.

Now we're going to read the important aspect here.

We're going to get into the meat of the matter.

34:49

Directed energy weapons, which destroy targets with energy transmitted at the speed of light over long distances are in a class of their own.

The other three weapon system types destroy targets by delivering mass to the target using either the kinetic energy of their own velocity and mass, or the stored chemical energy of conventional explosives to destroy the target.

35:13

Each type of weapon operates in different ways, is suitable for different kinds of targets, has different response times, and requires different numbers of weapons in orbit to achieve the degree of responsiveness required to reach a particular target when needed.

35:29

And then it says here that there's a table that summarizes these distinctions, but I think we summarized them pretty good.

So directed energy weapons are in a league of their own.

Let's keep that in mind.

Now, when we're talking about directed energy weapons, we're not just talking about your standard laser beam, which lasers can be quite powerful, don't get me wrong.

35:53

But we're talking about scalar wave technologies, folks, which are in a class of their own.

And I don't think it touches upon that aspect of it in this paper.

But these things most definitely do exist, and their potentials, the damage they can cause is enormous.

36:15

But let's read what they have here about directed energy weapons.

Now bear in mind this is from 2002, likely talking about technologies that were 30 years old at that time.

Keep that in the back of your mind as we go through this.

36:31

This is just what's acknowledged publicly 20 years ago.

Now imagine what is possible today.

Directed energy weapons.

Directed energy weapons include a range of weapons from electronic jammers to laser cutting torches.

36:51

Well, jammers need to transmit only enough power to compete with the targeted receivers intended signals.

Destroying ballistic missile boosters would require developing and deploying lasers with millions of watts of power directed by optics on the order of 10 meters in diameter.

37:10

Directed energy weapons could destroy targets on or above the Earth's surface.

Going to pause for a moment there, folks.

I'm going to repeat that sentence for you.

Directed energy weapons could destroy targets on or above the Earth's surface, depending on the wavelength of the energy propagated and the conditions of the atmosphere, including weather, including weather.

37:43

Although the energy a laser delivers propagates at the speed of light, the laser has to hold its beam on a target until energy accumulates to a destructive level at the target.

After destroying a target, it can retarget as quickly as it can point at the next missile.

38:00

Should it have sufficient fuel when defending against a salvo of missiles, the laser will only be able to destroy a certain number of missiles.

Well, they are in their vulnerable boost phase.

Going to pause for a moment here folks.

38:16

Now remember, they frame this up as a means of missile defense.

And this is what President Reagan's Star Wars initiative, or SDI initiative in the 1980s was about, these technologies.

Like I said, these were likely 30 years old at the release of this in 2002.

38:36

Bare minimum, maybe even older than that.

And that's of course the public face of what they would have these for.

But remember, it could destroy targets on or above the Earth's surface.

Directed energy weapons do exist.

38:55

They have existed for a very long time.

This is not speculation or conspiracy theory or tinfoil hat and utter nonsense.

They exist, they can be used to start massive fires.

And they have that wonderful aspect of plausible deniability built into them, don't they?

39:16

Especially if they're an alleged space based platform.

Something way up, high in the sky that people can't see sends an invisible beam down to destroy a target.

Something you can't see but has a most definite effect.

39:39

Let's go ahead and continue reading.

So it's talking about missiles in their vulnerable boost phase.

So the number will depend on the laser's distance from the launch position and the hardness of the missile target.

The farther the laser weapon is based from the target in, the harder the material of the target, the fewer missiles the laser will be able to destroy during boost phase.

40:01

Because the distance of laser satellites from missile launch points fluctuates in a predictable way, an opponent launching missiles will be able to choose to launch at times that allow the maximum number of missiles.

To penetrate the defense.

40:17

Going to pause for a moment here folks.

So most clearly it says right here in this Rand Corporation document laser satellites naming the very weapon system here that would make this possible.

40:36

Now if you can use a space based satellite with laser technology to destroy missiles in flight above the surface and it's been stated that you could destroy targets on the surface, well what do you think?

40:53

Especially if you can start, oh say a fire right as a windstorm is moving in, and then you go ahead and decide to do everything in your power to maximize the damage potential by mismanaging, grossly mismanaging the response to said fire, You have a recipe for psychological operations to go on.

41:23

It's a type of warfare.

It's real warfare.

It's domestic war, folks.

Domestic warfare, unacknowledged domestic warfare.

Governments harming their own population.

41:42

All that's missing is some incentive to do so.

The capability is certainly there.

Everything's present.

They've fulfilled the occult necessity to tell you what they're doing, garnering your metaphysical consent.

42:11

That's the term I use to acknowledge this, metaphysical consent.

They manufacture metaphysical consent by telling you what they're doing, by putting out this synchromystic metadata into the ether for people to pick up out there into the zeitgeist or the collective unconscious, if you would prefer to use that term to describe what this is.

42:33

This synchromystic metadata constitutes this information field.

If you prefer the term quantum information field, If you want to be more tech sounding or modern sounding, this is certainly something acknowledged.

It's the collective unconscious, it's the Akashic record, it's genetic memory, epigenetic memory, all of these things combined together, all of these different things that we call this phenomenon, the zeitgeist, if you will, doesn't matter what you call it, It's definitely something that's there.

43:08

It's an energetic field, it's an information field, and we can detect on an unconscious level some of these archetypes and tropes put out into this information field.

That's what I call the synchromystic metadata.

43:24

So that's been out there, thus preparing the way occultly for this.

Garnering your metaphysical consent through a contrived way, of course, but this is one of the ways that these dark occultists who run things in this world attempt to skirt karma.

43:41

They tell you what they're doing.

It's much like the old trope of the vampire.

He can't come in unless you invite him in.

It's the same thing with this type of thing on an occult level.

They need to get your tacit consent first.

How do they do that?

By subtly telling you what they're doing, giving you hint after hint after hint, and even in recent years, they hit you over the head with it and you still don't get it.

44:08

But in so doing, if they do that and you don't object to it in some way, shape or form, or call it out ahead of time, that gives them the tacit metaphysical consent to do those things they want.

And then they perform the operation and then they collect the results that they want.

44:26

This is how they operate things, and the mere fact of discovering what's been done after the fact empowers it all the more.

This is what's called revelation of the method.

It gives them a good laugh and chuckle at us.

That's why it's kind of a, I don't know, a mixed bag.

44:44

It's a Catch 22 whether to cover this type of stuff or not, because in covering it, it is giving the revelation of the method here.

But I think we need to demonstrate that this is a real thing.

So it's an important service to provide to people.

But at the same token, it's also kind of given these people a little bit larger of an ego because they think "we've just pulled another one off".

45:09

That's why it's important to look at this stuff, because we need to prevent more of this kind of thing in the future.

And if we could begin to recognize these signs out there, these synchromystic metadata in the zeitgeist before an event, we can head it off and it won't happen.

45:30

And their House of Cards crumbles and they can't pull off whatever it is they want to pull off.

They haven't pulled off another one over us.

That's why it's important to learn how to read this information field.

Because if we can read the information field and head it off at the pass, then perhaps we can prevent some future tragedy from happening or some future social engineering trope from happening.

45:57

And that's why I attempted earlier to go ahead and name the date of September 29th here.

And like I said, it is conjecture on my part.

I reserve the right to be totally wrong, and I hope I am wrong about that.

46:13

But around that time I suspect they're going to pull off some other grand event on the level of COVID.

And I'm hoping that if that is the case, if that is in the works, that is what they're planning based upon the things I see in this information field out there,

46:32

If that is the case, the mere mention of that here, if enough people start to recognize this and call this out and see it as being in the possibility here, then it won't happen.

They'll be forced to back out of their plans because if enough people are aware that something's coming, they can head it off at the pass.

46:54

And that is my hope in doing this kind of thing.

That's why I think it's important to demonstrate this.

But there's a lot of ground to cover.

That's the problem.

We need to frame this in a way that people can understand the concept and apply it logically to the things that they see and be able to piece it together and connect the dots and make some sense of it, and perhaps be able to better ascertain what future things may be coming based upon these things.

47:26

Not to think of it as silly or backward like we've been taught.

We've been indoctrinated too.

We've been taught to think that this type of way of thinking is archaic and backward or magical thinking or primitive, somehow stupid.

Magic doesn't exist.

47:44

Science.

Science is just like magic, only it's real.

How many times have you heard your science friends say something like that?

How many times have you heard that when they don't friggin have a clue about how things really operate?

48:02

There's an occult side to everything.

There's a metaphysical side to everything.

There's a spiritual component to everything that's largely been ignored by our modern society because that's all part of the zeitgeist they want.

That's steering us towards this notion that Rudolf Steiner called the spirit of Ahriman.

48:23

I call it Antichrist, keeping people bound in materialism,

hypermaterialism, so focused on the material aspects of things that they miss the broader, bigger picture here.

48:40

And I think that's what's happening with this whole event in Hawaii.

People are more focused on the nuts and bolts aspects of it.

What caused the fire?

Was it a directed energy weapon?

Well, regardless of what caused it, the fact of the matter is the more important component here is the spiritual component, the social engineering component.

49:03

What are they doing with this event and did they have something in mind ahead of time with it?

And I think we could demonstrate all day long that these occult concepts are definitely there, undergirding the whole thing.

49:24

Looking back and seeing this synchromystic metadata pointing the way to this, the patterns that we see emerge, recognizing the pattern, that's the more important portion of it.

49:43

So what happened after the major fires last time? We had lockdowns.

Watch out for the lockdowns, folks.

That's the important thing.

I could tell you here, maybe they're going to introduce a new germ on the world of some sort, some type of an invisible threat.

50:02

They've been talking about disease X in public, now talking about the next pandemic.

And of course, there's a new strain of COVID for the fall, of course, because they've got to put out a new vaccine, don't they?

50:19

All these things coming around again.

They got away with it before, too many people capitulated to the nonsense, and they're trying again.

I think they're going to get a little bit more push back at this time, so maybe they're going to amp it up a little this time.

50:37

Maybe they're going to make the stakes a little higher this time.

We'll see.

I hope to be totally wrong about all of this, but the pattern recognition in my soul,

I see the pattern, and I know they're up to something.

It's a repeating pattern of 2020.

50:55

Go back and carefully scrutinize the events that were in the news media in 2020 and the sequence of those events and align them with what's going on today in 2023 and you'll see a pattern begin to emerge.

51:13

Now, this time around, they've imbued this pattern with Eastern philosophies and ideas and energetic principles, Chinese based ideas and principles.

You see, they've given a strange amount of prominence and importance early on this year in the notion that this is the Chinese Year of the Rabbit.

51:39

Go back and look at the news stories circulating back in January, February, the early part of the year, Chinese New Year's celebrations here in America being shown on television all over the place for the first time I can remember in my 48 years of life, they've ever made a big deal out of Chinese New Year.

52:00

And we had the Chinese spy balloons, which very much look like the Chinese lanterns flying over the the country.

You remember all of that, The Year of the Rabbit?

They pushed the notion that it's the Year of the Rabbit quite a bit.

We have all of the symbology pointing to the rabbit and the moon.

52:21

You see in Chinese mythology, there's a moon rabbit and there's the whole mythological archetype that goes along with that.

You might have to go back and look at some of my other work to explore that topic more.

52:39

It's not something I discussed on the podcast all that much, but Rose and I covered it on the Poppycock Report, I recall.

And I do have some substack articles out there about it, but this whole Chinese New Year, Year of the Rabbit bit, there's something there.

52:58

They're leveraging some type of an archetypal component

with that.

I don't know what that means or what it's going to look like, but I do suspect based upon the importance that they've imbued on the Chinese New Year and the whole year of the Rabbit trope that this next upcoming festival this mid autumn or mid fall festival as it's called, if a memory serves me, that they celebrate in the Chinese culture is on September 29th.

53:25

So that's why if I had to pick a date that I think they would try something, it might be then.

And of course we are approaching the 22nd anniversary of 9/11, the number of the master builder, 22 years ago this happened.

53:44

So this is of massive, massive importance, what's the word I'm looking for, a milestone, I guess, for the occultists, 22 years.

So something's afoot.

54:02

And of course they always like to leverage the idea of the fall, because it's that time of year when energies, life energies, begin to diminish and start to drop off and are on the descendent then.

54:20

So they like to leverage things around the equinoxes because of this, the fall equinox in particular, because that's when the descension of the energetic principles begins.

And then in the grip of winter, when these energetic principles are at their lowest, that's when things really begin to get bad and amplify.

54:46

But the setup is always in the fall for many of these things.

There's other reasons I'm sure along with this, but these are just the basic premises here.

Now of course I keep harkening back to the occult aspect of this, but rest assured, for those of you that are more concerned with the material paradigm and the material side of this, directed energy weapons most certainly do exist and they can perform this function, and likely there are weaponized satellites.

55:22

Whatever you think a satellite may be, there's something up there in the sky that can fly over a target and do this most certainly, assuredly.

These things exist, so keep that in mind.

55:41

Now ostensibly they develop them for, quote-unquote missile defense.

And they do have other space based weapons that are kinetic based things that'll shoot an iron rod down to the earth with destructive capacities that are unbelievable.

55:58

I've seen tell of these as well, but getting back to the directed energy programs, the directed energy weapon programs, they've imbued them

also with mythological archetypes, they have names like Thor, Odin, Helios, I'm not kidding, you could look these up.

56:17

These projects exist, these technologies exist.

They're out there and they're all directed energy weapon technologies, no doubt about that.

But I wanted to go ahead and look again at some of the social engineering aspects of this whole thing.

56:43

So here, this is from USA TODAY, Today, this morning, 17 hours ago.

Here's the headline quote.

The next Maui could be anywhere.

Hawaii tragedy points to US wildfire vulnerability.

57:00

Published this morning, updated 5:08PM today Eastern Standard Time, August 19th, 2023.

The article says as follows.

The deadly wildfires in Maui reveal a vulnerability in the United States that is increasing as quickly as threats from climate change.

57:23

Huge swaths of the nation lie in dry danger zones where wildfires spark and cash strapped governments have ineffective emergency plans to save lives.

That was the deadly combination in the Maui disaster, namely wildfire risk, coupled with what some experts and victims have called questionable emergency preparedness.

57:44

And it has played out in some of the deadliest fires in the nation and around the globe, alarming fire experts and community leaders.

Similar scenarios happened in Paradise, CA, where 85 people died and nearly 19,000 structures were destroyed in the Camp Fire in 2018, and in Algeria, Italy and Greece, where questions of effective emergency response and preparedness have been raised after more than 40 people combined died from wildfires sparked by an intense heat wave, high winds and dry vegetation last month.

58:20

Canada is experiencing a devastating record wildfire season, with over 33.9 million acres scorched and at least four people dead so far. In Maui, where at least 111 people have died and more than 2200 acres were burned in the August 8th wildfires,

58:42

the county already knew it had a high wildfire risk, according to a study at Commission two years ago following an unprecedented wildfire season in 2019 where more than 20,000 acres were burned.

Hawaii's and Maui's fire problem is more extreme than on the US mainland, the study said, noting dozens of buildings and vehicles were damaged in a 2018 wildfire.

59:07

While there were no deaths in either of those years,

Warnings were raised and possibly not heeded by local officials.

Now, experts from around the world are taking a second look at many places that may also be at risk after Maui's crisis, which now is among the top 10 deadliest wildfires on record in the US since 1871.

59:29

The next Maui could be anywhere, said Turtha Banarji, a civil and environmental engineering professor at the University of California, Irvine.

Going to pause for a moment here, folks.

They always got to get their UC professor in there to speak, don't they?

59:45

Their expert on climate change, right?

What else did this Banarji say?

Quote

Realistically, almost any place could have a wildfire, End Quote.

America isn't the only country worried about wildfires.

1:00:04

Well, of course not, right?

Thousands of communities, from urban enclaves to coastal towns and remote locales throughout the US and abroad similar to Maui, are vulnerable to wildfires because of the increasingly deadly combination of climate change and government's lack of emergency plans and resources, experts say.

1:00:25

Going to pause for a moment, your folks, and call absolute poppycock.

They're always, always, always planning scenarios, emergency planning.

They throw money at it like there's no tomorrow.

But of course, corrupt politicians put that money right in their pockets rather than doing what they should be with that money, and that's a factor in this too.

1:00:46

So they want you to be programmed into thinking, OK, runaway climate change.

Things are getting so bad with the climate.

And by the way, our governments are so totally inept and mismanaged and absolutely, unbelievably, unbelievably corrupt and or incompetent that they can't plan for these things properly.

1:01:14

So this makes for a greater danger.

The climate change aspect of it certainly is nonsense, and I certainly think that if the governmental side of things here, the mismanagement and the incompetence factor, are really factors, well, that should be something that's usually concerning and telling because people are purposely being appointed into these positions knowing full well that they are incapable of performing the task.

1:01:48

So this is deliberate as well.

So let's go ahead and read on.

Now, of course you're science, a professor from UC, University of California Irvine says, quote, there seems to be a consensus among those in the scientific community that it might get worse for a bit before it gets better,

1:02:09

End Quote. Really? The consensus?

So climate change is going to get worse before it gets better.

Of course, if you pay enough carbon taxes, maybe it'll get better If you eat the bugs and, you know, freeze to death in the winter time because you don't use any heat source and drive your electric car, then maybe the climate will get better, right?

1:02:34

It's nonsense on the face of it, but I find it very telling here.

They're trying to prepare people to expect more of this, that has this smack of preplanning inherent in that too.

1:02:49

They want you to believe that this could happen anywhere.

These wildfires can just randomly crop up.

I mean, look at Canada's on fire, Hawaii's on fire.

We had that event in California.

Australia burned, you see, And it could happen.

1:03:12

And the reason it happens is because of that climate change and because of government, government ineptitude, government mismanagement, they don't have enough planning.

Do you see that trope being pushed here too?

1:03:31

I guess they want us to believe that things like Event 201 are good things when they plan these things.

They want to have responses planned out for this stuff.

1:03:47

They love their simulations, these people.

So I'm not buying it, folks.

I don't know if you are.

I don't buy it that the governments are ill prepared.

They're ill prepared for these things.

No, they're not.

They're doing what they're told, they're doing what they're instructed, and they're blindly following orders.

1:04:09

Just like that guy in Hawaii who didn't want to release the water, turn the water on.

Like the guy that didn't want to sound the alarm because it's only supposed to be sounded for tsunamis and he was afraid that if he sounded it, the people would run towards the fire and into the fire.

1:04:28

Really, no common sense, just blind order followers, sheep.

The same ones that I'm sure walked around Hawaii with the mask on, on the beach, distancing themselves 6 feet apart from one another.

1:04:46

Running out and getting their multiple vaccines because they were told they have to,

Being ostracized if they didn't.

And these are the types of people we put in charge.

1:05:02

Order followers, not leaders.

That's the problem.

You put order followers in the positions of leadership, the ones that should make decisions.

Responsible, common sense decisions that could save lives.

1:05:18

And yet they won't do it because they weren't told to do so.

They were told this is only for this.

Do not use this in any other occasion.

That's not leadership folks.

1:05:36

Leaders make decisions that sometimes go against whatever the rules of operation are or were out of necessity, out of common sense thinking.

I mean, it's common sense to think there's fires over there.

1:05:54

Maybe I should turn on the water so that the water is available there for them to fight the fire over there.

No, can't turn the water on.

Can't do it.

It's this lack of common sense.

And if this is actually the people we put in charge of things, I think we need to rethink what we're doing here.

1:06:18

Were these elected officials?

Probably not.

They were probably appointees.

Probably somebody's nephew needed a job, somebody's nephew that the governor knew or something needs a job.

Okay.

Well, Pat Skippy on the back here.

Go ahead.

1:06:33

You sound the alarm

If there's a tsunami there, Skippy Okay.

I'll do my job.

And that's what you get.

So is that how things are done?

I don't know.

Or is there more involved with this?

Or do we believe the story at all?

1:06:49

That's the other aspect of it.

I'm on the fence with a lot of it.

I don't know what to believe about this whole thing, but what I do know is it is a harkening symbol, signal for some other event to come afterwards.

1:07:06

It aligns perfectly with the timeline of 2020 with the sequence of events.

The pattern.

Like I said, it's about pattern recognition at this point, pattern recognition and being able to adapt the pattern to what the zeitgeist is at the time.

1:07:23

And if you're not comfortable with the term zeitgeist, which simply means spirit of the time, use the word collective unconscious.

If you would prefer use something more sciencey sounding, quantum information field, whatever you choose to call it, rest assured it exists and the unconscious human mind is very much aware of it and sees and recognizes these little bits of data in that field.

1:07:51

And it will affect you on a subconscious level that you don't usually consciously recognize, but it will affect your conscious behavior at a later time.

That's the whole key to all of it.

And that's why they manipulate this data field in that way.

This is how one of the ways in which they affect human consciousness, how they affect the public consciousness of people, how they steer mass behaviors.

1:08:17

It's mass psychology, folks.

At the core of psychology is occultism.

In case you were unaware of that, go back and listen to my show where I looked at the roots of modern psychology.

1:08:32

It's based in some of these occultic teachings.

It's based in these old ways of thinking.

That's where our science, or quote-unquote science of psychology comes from.

And much of this science that they call psychology is very much subjective.

1:08:48

There's no objective measures of it.

In fact, I just saw a study that these SSRI drugs that they prescribe all the time for depression likely do not work at all.

They've come to the conclusion that depression isn't caused by serotonin level discrepancies.

1:09:10

There's no evidence to support that, even though that's the operational theory that was put out many years ago and is still practiced by practicing psychologists or psychiatrists today.

When they prescribe medications for people, they assume depression is caused by some type of a serotonin level discrepancy.

1:09:30

So they give you an SSRI, a selective serotonin reuptake inhibitor in order to remedy this problem.

And most of the time it causes worse problems.

And they call this science, but they really don't know what the underlying mechanism is for this.

1:09:49

They assumed they came up with a reason.

Well, it must have something to do with serotonin levels, and that's what they applied as a standard of care for this.

And that's one of the things that goes on here, even though they really don't know what the mechanism of action is, have no frigging clue what causes depression, but they assume it's some physical thing, some physiological thing, and that's what's been done here.

1:10:19

They've attached physiological mechanisms that don't actually align with the reality.

Too many of these things, in order to instantiate some type of a control of the system.

This is cybernetics methodologies, Folks, I hope I'm not losing anybody here, but this is exactly what's been put into practice in this world today.

1:10:39

That's why they seek to quantify everything.

They want to try to attach some type of a material world cause and effect sequence to everything.

And they do that through quantification, count and measure everything.

Everything is a type of a particle interaction.

1:10:58

Now that doesn't necessarily hold true, But the thing is, even if it's not true, they could come up with approximations that come close to some of these things.

And that's what they've attempted to do here.

They come up with an approximation, a physical approximation of what they think could be the potential unit of measure for a thing, and this way they could instantiate some type of control over it.

1:11:25

Even if it's imprecise control, they can still steer things in certain directions.

That's what a lot of this is based on.

That's what cybernetics methodologies are.

You quantify it, and if you quantify it, if you could count it and measure it, you can better control it.

Then you could define a mechanism of action.

1:11:43

And like I said, even if it's not true or not correct, it gives you some tool that could steer things in a general direction.

So even if it's imprecise, it gives you an approximation of things that you can do with this mechanism.

1:11:59

And that's what's been done with many of the physiological sciences and especially the science of psychology.

So now what they've done is they've applied this in a way, in a particular type of medication to a particular type of condition.

1:12:15

And I'm talking of course about depression, which the studies, the new studies that have come out have shown serotonin levels have nothing to do with that.

But see, what they've done is they've looked at discrepancies in serotonin levels and they decided okay, well, we can apply this in a direction.

1:12:35

And so by applying these medications, what they have done in fact is cause depression.

These things cause greater depression.

That's one of the side effects that it tells you.

If you're depression worsens, you might need some other medication to help control this then.

So this is what they've done.

1:12:53

So now they've doubled down on the condition, they've engineered it into something physical or physiological that can now be controlled in this way.

Do you see how devious this really is?

And these doctors think they're doing good in this way.

What they've done is they've found this condition which they've identified and they've diagnosed in this person, and they apply this medication to that thinking that this will be the fix.

1:13:22

But what this does in fact is it reproduces the condition all the more, but it reproduces it on some physiological basis.

And this creates the feedback loop that they need in the system to control the system.

This creates that circuit.

1:13:41

This is called a causal circuit.

So what they do is they insert this causal circuit and now they've induced this condition even more manifest into the person in a physiological means.

And now that mechanism of action is present in the physiological system and now they can manipulate it in certain ways.

1:14:03

So this is a lot of what's been done through this system of psychology, this science of psychiatry or psychology, Physiology.

These things have been measured out in this way and had these cybernetics principles applied to them, because cybernetics is the study and the science of systems control, whole systems control.

1:14:25

So it's all about controlling the person's mind and behaviors.

That's what it's about.

So it's about inputs and outputs.

So the input you give is, I'm feeling down, I'm feeling depressed.

They diagnose you with depression and they give you this drug that ostensibly will help make you feel better from that depression.

1:14:46

But what it actually does is oftentimes it will further ingratiate that depression into you over the course of time.

You may have a positive response at first, but over time that degrades.

1:15:02

Over a very short time, most of the time, that degrades.

So that's one of the things that happens.

So in introducing this, they take control of the physiological mechanism, the triggering mechanism, and then they could apply other standards to it and control that, control somebody's behaviors, control their attitudes, control their emotional states on some level or another through the use of medications primarily.

1:15:35

But this is just one example of how they do things.

And I know that seems like a side trail from what we're talking about, and certainly it is, but it certainly applies very much to the reality we live in.

So remember that a lot of this has to do with the spiritual component, or the lack thereof, that they're trying to engineer into society.

1:16:01

It's all about keeping people focused on the materialist paradigm.

So of course we're more concerned when it comes to events like this with the why the how, What happened exactly?

How did this come about?

How did this come about?

1:16:23

We're more concerned with the physical cause and effect notion of things in these regards when we should be looking at the bigger picture.

What is it that's being achieved through this?

One of the easiest questions you can ask yourself about an event like this is who benefits from it?

1:16:45

Who benefits from this?

Well, I think we know the answer.

There's large real estate developers and government contracts and projects that have cropped up over this that may wind up buying some of this land for pennies on the dollar and rebuilding.

1:17:07

I think that stands to reason.

So that's who benefits.

So if you look at who benefits from it, you might be able to ascertain who is behind the event, who at the very least allowed this event to happen and capitalized on it.

1:17:29

It really is that simple when it comes down to it, if you want to ascertain who it is on the material paradigm of things, that's responsible.

But it goes much deeper than that, folks.

It's all a spiritual battle at the heart of all of it.

1:17:47

And this event, whether you believe it happened the way it's reported or not, it's irrelevant.

It's out there.

It's taking up human mind cycles.

Right now, people are talking about it.

1:18:04

It's a major focus, and it has us focused on this primary event and it has us focused on these concerns.

Well, what happened?

What caused this fire?

Why did these people stand down and let this happen?

1:18:19

Who stands to benefit from it?

You could explain all of this in the materialist way, of course.

You could say it's all about money.

It's all about money.

And that's what the political explanation may be.

1:18:39

But if you focus only on that, you're missing the bigger picture.

And that's the important part here.

It's a grander spiritual battle.

That's what's going on.

It's an attack against the human psyche.

1:18:56

And this is only one bit of information in this information field, in this pattern of information that's been presented that shows the roadmap of what's coming.

So we need to not lose focus on that aspect of things because there's something else coming.

1:19:19

Now, if these people in Hawaii really did suffer in this way and there was property damage and loss of life, that's tragic and that's horrible and I feel for these people, I do.

But we need to not lose focus on the bigger picture because they have bigger things planned.

1:19:37

It seems to me much of this has been engineered in a psychological warfare type campaign, keeping people concerned and infighting about the wrong aspects of this.

There's not too many people who I've seen in social media who aren't questioning this event.

1:19:59

It seems to me that those people who in years and times passed would be skeptical of something as conspiracy theory sounding as directed energy weapons,

They are pretty much accepting that notion of things at this point.

1:20:15

So.

Has this whole thing been engineered in a way where it's keeping us in this state of anger about this?

Is it truly a revelation of the method, or have they really revealed, Yeah, yeah, we do scumbaggy things like this.

1:20:38

What are you going to do about it?

That's kind of the connotation you get.

People acknowledge they did this, well,

You know what the problem with that way of thinking is?

It gives these people that are new to this type of thing, new to this type of research, new to this type of way of thinking.

1:20:58

It gives them the impression that these people, these dark occultists who run things, are the ones who perform these psychological operations are all powerful and they're so far ahead and they have every contingency planned.

And that's not the case.

That's not the case.

1:21:16

They are not.

My job here, folks, is to break the algorithms.

That's what my goal is, to break their algorithms, to see through the lies that they hand us, to think outside the box and not fall into the Hegelian paradigm that they've handed us.

1:21:39

Because of course, everything has got two sides.

Every story has two sides.

That's not the case, folks.

There's more than two sides to every story, but they keep you focused on the two sides.

So you have the ones that will say, oh, it's just tragic.

1:21:56

It was a wildfire caused by climate change.

And these folks that were entrusted in government, well, they just made some poor decisions and the fact that it played out the way it did is just coincidence.

And then you have the other side that says this was deliberately started.

1:22:12

It was Direct Energy Weapons and this and that.

And you have these real estate developers going to be making massive money on it.

So these are the two sides.

But I'm here to tell you, step outside the box for a moment.

That's right where they want you.

They want you to be outraged by this.

1:22:30

And if people, real people, really did die and real homes were burned to the ground, that is tragic for sure.

But don't lose sight of the bigger picture.

That's right where they want you.

They want you to step in one of those two boxes.

1:22:47

You need to step outside the box, Look at the bigger picture.

Look beyond the material paradigm, the material cause and effect paradigm.

There's spiritual forces battling in this world today.

1:23:07

We're living in the age of deception, the age of deceit.

It's hard to know what's true and what's false.

And remember, primarily the places we're hearing about this story are through where?

Well, the mainstream media outlets.

1:23:24

How much do you trust them?

That's all controlled and contrived too.

They could present us with a story that may not have any truth to it whatsoever, could be a fictionalized, fictionalized account of things, we wouldn't know any different because the messaging is consistent across all the platforms.

1:23:45

And remember the narrative?

The news media, all media,

about 95 to 98% of all media is owned by no more

than six companies, owned by 6 companies or less.

1:24:06

All the major media conglomerates, and they all have centralized messaging disseminated from newswire services.

A handful of newswire services.

So all it takes is one person at the top of the power structure of the newswire service to send out a message and all of the news media will carry it and present it to you as if it is true, as if it's truth.

1:24:33

And they'll repeat, verbatim, word for word, this script they've been handed with their own local flair, as has been demonstrated in many videos.

And people accept it as true because it comes from the news.

1:24:52

And even if you see through it, it still has some air of credibility to you.

Just because of how deeply indoctrinated we've been in this world with this, That's problematic too.

1:25:11

So you don't know what to believe.

We don't know what's true and what's false.

And that's why they keep us in these little boxes.

Pick a side.

They give you one of two sides to choose.

And if you fall in line on either of those sides, you're right in their trap.

1:25:29

Because the box, folks, is a trap.

Step outside the box.

Look at these things from all angles and most especially the spiritual angle.

Pay attention to the occult side of these things because it's very important, these people utilize it all the time.

1:25:54

I always caution people whether you believe any of that stuff or not is irrelevant.

What you need to understand is there are people in positions of power in this world that very much believe in these things, and the things they do to act upon their beliefs will affect all of us.

So you may think it sounds silly, or a bridge too far to think in terms that there's planning phases of this wherein things begin to manifest years ahead of time, in something as silly as a Disney movie, pre-echoing future events.

1:26:28

But I assure you, it is.

There you go.

Look for yourself.

If you know the language of symbology especially, you'll be able to pick things out.

If you have more than a 30 second attention span and you could remember back a ways, you'll begin to see things.

1:26:52

So something as simple as, hey, here's this soda released in 2020 called Maui Burst.

Maui Burst by Mountain Dew, DEW.

You see that and it triggers something in your mind.

1:27:11

When you see the event that came to pass.

Then you understand a little something about this notion of Synchromystic metadata, as I like to call it.

I think we need to really apply that term to it.

It's a good descriptor, Definitely a good descriptor.

1:27:32

And you could connect the dots between this stuff and like I said, go back and watch that Disney movie,

Moana with Dwayne the Rock Johnson as Maui.

Watch the end sequence where the island is on fire, but it will be reborn at the end. Through fire,

1:28:01

All nature is perfectly renewed.

INRI, the alchemical concept.

Here INRI.

That's what the alchemists and the secret schools claim that that phrase, those initials on the cross of Jesus Christ, stood for, through fire, all nature is perfectly renewed.

1:28:24

They are the philosophers of Fire, after all, and this is part of their belief system.

So whether you believe it or not, like I said, it's irrelevant.

You need to understand that the people in positions of power in this world, that plan these things and begin to manifest these things in certain ways.

1:28:44

They believe this stuff and what they do to act on those beliefs will affect you whether you like it or not.

So it's best that you understand what motivates them, what it is they believe, why they do the things they do, and how they do the things they do.

1:29:01

Because there are ways you're being manipulated that you don't even know about, that you are unaware of.

And it has everything to do with these tools that I've laid out here, this notion of synchromystic metadata and how it applies to the collective unconscious or the zeitgeist, the spirit of the time.

1:29:22

Do you think it's a silly notion?

Look at what science says about it.

If you want to believe the scientific paradigm, this is certainly an acknowledged thing.

It exists.

Think about something like the 100th monkey effect.

1:29:39

How does that work?

Well, this could only work if there did exist something like the collective unconscious or the zeitgeist.

This is how memes work, folks.

This is meme magic, if you want to call it that.

1:29:56

It's the alchemical meme.

The use of synchromystic metadata to influence human behavior on an unconscious or subconscious level.

1:30:13

That is meme magic.

That is what this is all about.

That is how we are affected.

That's why Internet memes are such a powerful tool for awakening people as well.

1:30:31

That's why even things, even things like humor, humor, parody, all of these type things have their use and are powerful tools as well.

1:30:53

If you could laugh at something, then it doesn't affect you as much as it would otherwise.

So we need to laugh at these people, at their nefarious plans as they sit around in their little meeting rooms planning the future for those of us,

1:31:14

The rest of us, yes, we'll get them to eat the bugs.

If you can't laugh at that crap and that nonsense, then something is wrong.

They have more control over you than you would think then.

So if you laugh at their stupid notions, the less power that has to come to manifestation in this world.

1:31:35

So it's important that we use things like satire, humor, parody and memes to battle back, even though it's just little small victories here and there.

That's one thing that the computer algorithms haven't quite gotten a hang of yet, is being able to understand memes.

1:31:58

That's a tool and a resource that we need to keep using to awaken minds because they're working.

And because of this notion, no machine can emulate the human imagination.

1:32:15

No machine can understand the synchromystic, and that is where their big problem comes into play.

Their artificial intelligence just doesn't grasp it, even still, nor will it ever fully grasp it.

1:32:31

It doesn't read the language of symbology, and we need to leverage that to our own advantage.

And the day that we do that is the day that these dark occultists who run things in this world fear the most.

Because the very tool they designed to enslave us can be used to free us.

1:32:53

We need to fight back.

It's a spiritual battle, folks, with very material world tools in play, don't get me wrong, but these are small things we can do.

Break the algorithms.

That's wherein they have their stranglehold of power, the data set, they collect the big data, and they have data points on all of us, and they can manipulate and use these in many ways.

1:33:21

Of course, this is the cybernetics methodology, but they still rely upon the computers, the algorithms, the artificial intelligences to flag things for them that don't belong.

And if you are clever enough and you learn how to use the symbols and the various tools, the synchromystic tools against the system, you can break the algorithm and we can have a larger effect.

1:33:55

And I think we need to do that.

And I think memes are an important template to do that with.

So keep meming out there if you will.

1:34:11

Keep on going, man.

It certainly it has an effect.

Anyway.

The thing is here, the whole point tonight is, yes, Direct Energy Weapons do exist, and perhaps this did have something to do with these events.

1:34:30

But the more important take away is to recognize the broader pattern and know there's something bigger on the horizon here.

This is just one symptom, one particular piece of data in the grander data set in the pattern that you can recognize and be able to tell what may be coming next.

1:34:52

And then if enough of us can call it out, then maybe it won't happen.

That's the whole notion here, so take that how you will.

So I always caution you take a lot of this stuff with a grain of salt.

1:35:12

And I did tell you, and I will continue to tell you, that it's conjecture on my part.

I reserve the right to be totally wrong, and I hope I am wrong, but it seems to me the pattern has emerged and I do recognize the patterns.

1:35:28

It's following the same template as before three years ago, so take that with a grain of salt.

You can do with this warning what you see fit.

Whether you choose to ignore these things or to act upon them, it's up to you.

1:35:47

But the more of us that act upon them and try to spread these ideas and think outside the box, the better off we will be.

At least that's my opinion.

But if you're content to stay in the box, that's your prerogative.

I'm not going to tell you what to do.

1:36:05

I'm just giving you my insights.

That's all I can do.

That's all I can do.

And I'm a fallible human being as well, so I could be wrong.

Remember that.

1:36:20

Do your own research.

Look at these things.

Apply your own standards of logic and reasoning to them.

And the most important thing, folks, Keep your spirit and your heart and your soul right with God.

Get in that relationship with God the Creator.

1:36:43

We're living in unprecedented times.

We're seeing things unfold that have been predicted in the holy books, especially the Holy Bible.

Get right with God.

Time's drawing short.

I don't know what the future looks like, but I do know we've been given some warnings and we need to heed those warnings if we want to have a brighter future than what the technocrats have planned for us.

1:37:15

So keep that in mind.

It's a spiritual battle.

Get your heart right with God and it'll be all good.

That's the thing.

He'll never leave you, nor forsake you.

He'll be there.

He'll make a way where there's none.

He's demonstrated that in my life many times, so I want to offer you that encouragement.

1:37:38

So this is what we can do.

We could be in right relation with God, our creator, and we can fight back by recognizing the synchromystic, and also by using the subtle tools of memes, humor, satire, to fight back against this stuff because it disempowers these people when we do that.

1:38:10

And we need to do that on a grander scale.

Why do you think memes and stuff get you banned off of Facebook and YouTube and everything else?

They know the power in the meme.

Use it.

We need to use this to our advantage.

But anyway, folks, that's all I have for tonight.

1:38:26

I want to thank you all for tuning in.

Have a good night.

We'll catch you next time.

Have a good one.

Visit www.alchemicaltechrevolution.com to further support my work. Thank you and God bless you all.

www.ingramcontent.com/pod-product-compliance
Lightning Source LLC
Chambersburg PA
CBHW062350290526
45794CB00005B/2160